Random Thoughts

About Life And Other Insignificant Stuff

By Cliff Woffenden

Typesetting and graphics by the author.

Published by Howling Moon Productions
Box 423
Nakusp, B.C.
V0G 1R0

ISBN: 978-0-9694585-9-3

Cover photo by Carla Nemiroff (2014)

Dedicated to all those who thought they knew what was going on....
Surprise!!!

Intro

Beginning my journey

In my late teens and early 20s I rejected the bible, god and religion in general. Then I was introduced to the Moody Blues. Their lyrics and music inspired me to start my spiritual journey. I became ravenous to devour all I could concerning spirituality and our relationship with Life, the Universe and the Creative Force behind it all.

What I found was so far beyond anything I was taught or imagined. I looked at the world and saw that 2000 years of Christianity had got us nowhere! We were still barbarians, still killing each other over doctrine and dogma. There was still suffering, starvation, and war. Christians had learned nothing from the teachings of Jesus or the bible. Everything in this religion is about personal salvation and everyone else be damned. So I looked into the teachings of all the other religions I could find.

I spoke to numerous people of a wide variety of religious persuasions. I went to prayer meetings, healing circles, ceremonies and rituals, always looking for what would bind us all in spiritual union. What I found was that religion is the singular most disruptive force on Earth that separated us, each faction thinking that their way was the only way.

I eventually found a unifying principle that does bind us all as humans. It has nothing to do with religion or beliefs. It is only about loving ourselves as we would like others to love us. Love is the only unifying principle in the Universe. Religion does nothing toward this end. It only promotes exclusivity.

There is no faith in religion, only belief.

There has always been those who found faith through the path of religion, but they are few and far between. Faith can only be found with an open mind and heart. Thinking that you have found the answer is a sure way to stunt your spiritual growth, which is an evolutionary process that never ends. Even the Creative Force evolves as we do. God/Great Spirit/All That Is, is not a stagnant being and it is not separate from us or any other part of the Universe. It is everything and grows with everything in It.

Spirituality or Religion?

"Religion is for those who are afraid of going to hell, spirituality is for those who have already been there." - Robert G. Ingersoll

Practicing spirituality is not necessarily dependent on deities. It is a personal journey into the mysteries of life. My journey has taken me beyond religion, beliefs and deities. Although I believe in the possibility of some divine creator, it is so far beyond our comprehension that to dwell on it is a waste of time. What is more interesting to me is the divine nature of humanity. It is much closer to home and is available.

The Universe is infinite and we are finite. But we are made up of the same stuff the universe is made of: the creative energy that caused all this to come into being. As Jesus supposedly said, once we realize our full potential (our divine nature) we will be able to do far more than he could ever do. Jesus was trying to show us our potential, but those who would control us have worked diligently to keep our divine nature from us, otherwise how could they control us? Religion has always been a tool for control and the controllers wrote the bible.

A New Paradigm

I've been thinking. There seems to be two different things going on in the world. One is the death of the old paradigm with its penchant for power, control and violence. The other is an awakening to a new paradigm of Love and Harmony with all living creatures.

Of course I'm not talking about romantic love, that is more like hormonal overload. I'm talking about the love that permeates all of creation. Love is a state of being connected to Source, being in your center, acknowledging your own divinity. We cannot truly love another but we can share our love with another or with others.

We are coming into this new paradigm as children. It is all new but many of us have been preparing for it for decades. It is not something that we can understand with our minds. It is a heart based knowing; a knowing of the soul. I think this year might be a good one to celebrate our rebirth. Party on dudes and dudettes!

Who was the Word given to?

For some fifteen hundred years, the natives of the Americas never heard of Jesus or the god of the bible. They had no way of knowing about it. Those that say the only way into heaven is through the professed belief in JC as your personal saviour have condemned all people, who never heard of the word, to hell. That is the good news they want to peddle.

As one Inuit elder put it to a priest who was trying to convert him, "If I never heard of the bible, would I still go to hell?" The priest answered, "I guess not." To which the elder replied, "Then why the hell did you tell me?"

If they believe in a loving and kind creator, then why is it so difficult to see that the word was given to all people, in all geographical locations in every historical time frame? The bible is the account of the word given to Jews 2000 years ago, according to their understanding. The Australian Aborigines, the Chinese, the native Americans, the Africans were all given the word according to their understanding, or god is not a loving god. Period.

Setting Sail

Religion was created out of fear by promoting fear. It is quite obvious, that many Christians, particularly the born again evangelicals, become religious because they fear their own nature. They try to exorcise their own demons and try to find some inner peace from the constant battle in their hearts and minds. I understand the process and the healing powers religion can bring to the chaos and inner turmoil for those who have felt lost. But religion, in this case, is not the be all and end all that proponents think it is.

It is just the beginning of the journey. It is an anchor in safe waters. To begin a journey one must hoist the anchor and set sail. This is what Allan Watts meant by belief is holding to a rock, and faith is learning how to swim in the stream of life. I see religion as the first port-of-call on a journey to oneness with all of creation/god/universe or whatever. But first you have to leave port/move beyond belief. Accept who you are (a child of god) warts and all, not as a sinner but a brave explorer on a journey of discovery. The payoff is you finally get to know who you are without fear.

The Human Ego knows no bounds

I believe that all creatures big and small are the Creator's children. I feel no differently when I hear of a deer killed by a car than I do when I hear of a human killed by a car (or any other instrument of destruction). When I see a clear-cut forest, I get a physical pain in the pit of my stomach. I do not feel any differently about the death or wonton slaughter of any creature: wolf, bear or buffalo culls or wars on humans.

Humans have been heading downhill ever since the evolution of agriculture. It was agriculture that spawned the first civilizations and religions. It spawned an attitude of superiority, not only over nature but of those humans who were "not civilized" and made both targets for exploitation and annihilation. The Pope declared in the 1480s that any group of humanoids not mentioned in the bible could not possibly be human so it was perfectly okay to kill or enslave them and take their lands in the name of god. These decrees were called Papal Bulls and are still in effect today. It is that justification which allowed Europeans to colonize the world and to slaughter hundreds of millions of aboriginal people the world over.

It is this same attitude that motivates humans to think they have the right to "manage" nature to our liking, cutting down vast swaths of forest, extinct thousands of species of fish and wildlife. The human ego knows no bounds in its race toward the eradication of other species, not realizing that we are only a part of a greater whole and that the process will eventually lead to our own extinction. We are the only species that is hell bent on destroying the world's life support system, the biosphere. That is why I have no more sympathy for human suffering than I do for the butterfly that ends up squashed on the grill of my car. I see all life as having an equal right to life.

We are Stardust

One day, a long time ago, it came to me in a flash of inspiration, that the second coming was a process, not an event. Jesus is a metaphor, an archetype of a consciousness that has been around for thousands of years, even before JC was supposed to have materialized 2000 years ago. The ancients were aware about many things but they couched them in symbolism, with each culture using symbols that made sense to them. The Jesus story is as old as the hills and twice as dusty but using different names.

Over many millenniums, humans have evolved and we are fast approaching a state of consciousness that will allow us to understand what the metaphor was all about: we are no different than Jesus. We are all children of god/universe/great spirit or whatever. Jesus supposedly said that when we finally wake up to who we really are we would be able to do far more than he was able to do. He never put himself above others and never claimed to be greater than anyone else. It was the apostle Paul and Emperor Constantine who made him a god. They missed the point of his teachings!

We are all spirit infused in matter. We have the divine spark in us. Once we wake up to our true potential as human beings we will come into our full power. Religion is the biggest drawback to us fully developing our full potential by telling us that we are lesser beings, sinful beings, hiding our divinity from us.

"We are stardust, we are golden and we've got to get ourselves back to the garden." – Joni Mitchell

Your Truth is Just an Opinion to the Next Person

All anyone can talk about is themselves. You can think you are talking about someone else, but in the end, it is all about what you think and believe. I learned a long time ago, that what I know as true is just an opinion to someone else. That is what any of us has to offer the world: an opinion. It is a matter of integrity to speak our truth but it is only our truth, not the only truth. It really doesn't matter what you believe in, or don't believe in, because it is only relevant to your life. Just because you belong to a group who believes the same as you, doesn't mean it is relevant or good for everyone. For some reason I keep going back to the Monkey Tribe in Jungle Book by Kipling. We all say it is true so it must be true. Kipling was a wise guy.

Were Humans Created or Are We Part of Something Greater?

Do we owe our existence to some invisible god who dwells in the sky? How I see it, is that there may have been some form of consciousness that set in motion a series of events that cause the Universe to unfold and evolve, but is it aware of itself, the Earth and its' inhabitants? I doubt it.

When I look at the structure of a solar system, I see an atomic structure and when I look at a galaxy, I see a molecular structure. When I hear that the Universe is made up of billions, possibly trillions of galaxies, I think, perhaps what we are viewing here, with our very limited capacity, is the parts of the body of the Universe. Perhaps the Universe is a conscious being and perhaps that consciousness created itself.

The Earth is like a proton in orbit around a nucleus (the Sun). The Earth is but one of trillions of planets (protons) in orbit around trillions of suns (nuclei). When you begin to think on this level, you begin to see how incredibly insignificant this planet is to the overall body of the Universe. And we live on it. Do we really think we are the only ones living on a proton in this infinite Universe? I doubt it!

Humans are the Who that Horton hears. We are like quarks and quarks are so small that they have not even been observed yet. As far as I know, Quarks are a mathematical theory. Are you aware of every atom in your body? Not likely. So, it can be assumed, we are not aware of our quarks either. Is it logical to assume that the Universe is aware of all the Quarks in its body? I think to believe that is completely illogical, and yet there are those who believe that the Universe was created and that that creator is aware of them, favours them above all other sentient being in this infinite Universe. They even believe that it communicates with them. Hmmm! I will allow you to draw your own conclusion about what I think of that.

Is the Earth our Mother?

We can only speculate how life began on Earth, but here we are, living on it. How did that happen? Were we created here or did we evolve from the primordial soup? Are we the epitome of evolution? Or, are we just a part of the evolutionary process of the Earth? What is the source of life on this planet?

Well, I think you already know what I think of the creation theory. The aboriginal peoples refer to the Creator as The Great Mystery for a reason; the infinite is incomprehensible to our finite minds. But the biosphere that envelops this rock is alive and probably has a consciousness far beyond our comprehension too. It is a single living organism that we are a part of, much like a blood cell is a part of a body.

A blood cell is a living cell, but if we die, the blood cell dies with us. Therefore, we can postulate that the blood cell gets its life force from our living body. Likewise, our body gets its life force from the biosphere. This is what the aboriginal peoples understood, when they referred to the Earth as our Mother.

For our blood cells to be healthy, we need to make sure our bodies are healthy. For our bodies to be healthy, we not only have to take care of it, but we would be remiss if we did not take care of the body, of which we are a part, of the Earth Mother. Ooops! Does anybody think that we have done a good job of that? Could this be why we have such wide spread disease and suffering among "civilized" humans?

Cancer, respiratory and heart diseases, to name a few, are first world problems. Anybody who doesn't think there is a correlation here is just not paying attention. There is not much that an individual can do to change the attitude of others, but we can make an effort, on an individual or collective level, to change how we treat our environment.

Yes, there are those who spend a lot of time talking about saving the Earth, but talk is cheap. There is nothing we can do to save the Earth. It will save itself. People should think more about taking care of themselves. That is something that they can do, something they can influence.

What are thoughts and are they ours?

Are we thinking our own thoughts or are we just part of an elaborate computer program? The brain is a computer. It records and stores data. But, on the quantum level there is no matter, so everything is just pure energy. The Universe is just pure energy and yet consciousness exists, on a collective (the Borg) and individual level (your consciousness). So, to me, it stands to reason, that what people call god is the sum total of all consciousness and energy that is the Universe. Nothing exists outside this energy field.

Physical reality is just a program, like the Matrix. We are told (programmed) from birth what this reality is all about. Eventually we forget the program we came from, the one where we were just pure conscious energy. But everybody has been subjected to this program so we all co-create this reality. Buddhist say that this reality is just an illusion, like a dream, and I say that what we call the dream state is our true reality. This, to me, is consistent with quantum physics and mechanics.

Thoughts are part of the collective process. Some people say we are spirit having a physical experience and others say we are god having a human experience. But breaking it down to the quantum level, they are actually saying we are energy experiencing a (consciously constructed) material reality. Thoughts come from the collective consciousness and that consciousness is the sum total of all life on this planet (different planets throughout the Universe have a different set of rules that they exist by, which is why we may never discover life on other planets - they may exist on a different plain of reality).

When we scan the Big Picture, we find there is no such thing as a personal god but some form of consciousness (beyond our comprehension) may have programmed the whole deliriously magnificent dance we call life. I can only speculate that it is the Universe itself.

Enlightenment

What does it mean to be spiritually enlightened? In the context of the dominant society, is it even possible? What happens if you do become enlightened?

It is not possible to become enlightened by seriously pursuing it through some discipline or practice like meditation or yoga, vision quest or sweat lodge. Why? Because spiritual enlightenment means a lightness of spirit.

You can't be serious!?

No, you can't.

The denseness of physical reality is too heavy for enlightenment. We are light beings, first and foremost; conscious energy. In the end, we will shed our earthly body, by necessity, in order to become enlightened and return to our pure form. I do not think very many people are willing to give up their earthly vehicle to reach the state of pure conscious energy.

The dominant culture is all about ego; the satisfaction of the physical needs of the ego and the body.

I hear a lot of talk about spiritual development and spiritual practice but what I really hear is "my ego thinks we are progressing because we know all these prayers, songs and spiritual information". I don't use the word knowledge here because it is mostly information.

Knowledge comes with integrating that information into the way we interact with others, with compassion and empathy. Information is useless unless it is shifted from the mind (ego) to the heart, the true seat of wisdom. And, you cannot fake wisdom or compassion.

If you expect a reward for your acts of kindness, your sympathy with those less fortunate or suffering you are practicing bartering or commerce. Unless you can be truly selfless in your actions, you are being fake; you are coming from ego.

If you are not willing to let go of your earthly attachments, your body and your mind, you are not ready for enlightenment. You are not ready to go home (to return to being a conscious spiritual being). When it comes down to it, being spiritual is about accepting that you are a spiritual being at your core and that this body you occupy is just a borrowed vehicle to navigate through the density of physical reality.

And really, physical reality is a construct, a Matrix that humans created because they forgot who and what they truly are. Those who would control us have denied us our divinity. We really are a species with amnesia.

The hundredth monkey

The truth of our origins is actually clouded in mystery. We have purposely been dumbed down. We are a species with amnesia because if we knew the truth of who we are and what we are, we would not need overlords, rulers or governments. We are programmed into thinking this "reality" we live in is the real thing. It isn't. It is a lie, a veil that has been pulled over our eyes to keep us confused, manipulated and compliant to our lords and masters. The truth will become self-evident very shortly. Some of us are already aware and we are waiting for the rest of humanity to reach "the hundredth monkey".

The Hundredth Monkey theory is that once a critical mass of a species reach a certain stage of development, the rest of the species will evolve to the same state. It is based on scientists' observation of monkeys living on a remote island off of Japan. One scientist taught a young monkey to wash its yams before eating it. It taught its mother and soon the whole tribe, on that island, was washing their yams. It was then discovered that all the monkeys of the same species on remote islands began to wash their yams, even though they had no physical contact.

God

What does the word god represent? For most people it represents an omnipotent entity that created the Universe but is not part of it, which takes a personal interest in our lives. That entity does not exist. It is pure fabrication by an immature, insecure ego that cannot come to terms with its own insignificance.

Let me put it this way:

Horton Hears a Who by Dr, Seuss. Horton was an elephant who hears a voice coming from a spec of dust floating by. He eventually discovers that an entire civilization occupies this tiny spec of dust and they are calling out to see if there are any other life forms out there occupying other specs of dust. Does this sound familiar? We are the "Who" that Horton hears.

Horton is an entity that occupies a world with a multitude of other entities and this world is just a spec of dust to other worlds occupied by a multitude of other entities... ad infinitum. There is no way that the Who can comprehend who or what Horton is. In our terms, the Who would think that Horton was a god. But Horton did not create anything nor is he any more divine than any other entity.

Taken as a metaphor, the story parallels our own. We occupy a tiny spec of dust in an infinite sea of dust and we are trying to contact other specs of dust (SETI) to see if we are alone or not in the infinite sea. Our comprehension of the vastness of this sea is infinitesimal. We call this sea the Universe and nothing in it exists outside or separate from it. There may be other universes out there but that is a speculation that is of no use to this discussion.

The Domestication of Humans

From the day we are born we are told what to believe, how to behave, what is acceptable behaviour, what is real, what is right and what is wrong, what is beautiful and what is ugly. We accept these "truths" because we are too undeveloped intellectually to discern the difference. We are too dependent on our caregivers, parents, teachers and priests or ministers to question the validity of what we are told. We accept a belief system without question and those beliefs manifest in our lives as reality.

We do not choose these beliefs or this reality. They are given to us. They are what we all have been given and it forms our collective reality because we have all been indoctrinated into the same set of beliefs. We agreed to accept the reality we were given. What choice did we have? There were no alternatives offered.

Humans have been domesticated through a process of repetitive reinforcement, through reward and punishment. This system of domestication is no different than how we train a dog or any other domestic animal. Religion and governmental laws reinforce this system of reward and punishment. Break the rules and we are punished. Obey the rules and we are rewarded.

The reward we get is positive attention from our parents, teachers and peers. The attention feels good, so we become addicted to it. We crave more, so we do all we can to get more. We start acting. We start pretending to be someone we are not in order to get attention and soon we become what others expect us to be. But it is not who we are. We become so removed from who we are, that we forget who that is. Good dog!

The domestication process is so strong that we eventually do not need outside influence to continue the process. We domesticate ourselves. We reward ourselves when we obey the rules and we punish ourselves when we disobey the rules we have been indoctrinated into. And we judge others by how well they adhere to the rules.

This is how we become socially conditioned by the society around us to be good citizens. There is no freedom in this, no self-actualization, no self-awareness; just domesticated automatons. Mostly we go through life on automatic pilot, going through the motions without really questioning the validity of what we bought into. To question the law, the belief system, is to put oneself on shaky ground, socially.

Because the system of reward and punishment is based on fear - fear of not being good enough, smart enough and loveable enough - questioning the rules makes us feel uneasy. Even though we may intellectually see there is something wrong with the rules, we have bought into them for so long that they have a stranglehold on our psyche.

If we do manage to change a belief later on in life, there is still a residual impact of the original belief. That is why it is almost impossible to try to change someone else's beliefs. Arguing is an attempt to change someone else's belief about some aspect of this "reality". Rarely, if ever, does anybody change his or her belief even in the face of overwhelming evidence. The program is too strong, too well encrypted.

Over the years, we collect information and experiences and that information and those experiences expand our belief system about life. We judge everything in our lives, people, events and the weather, by this system of beliefs. Nobody or thing is immune to our judgments, including ourselves. The only thing that separates each individual from the next is that we each have collected a different set of data and thus, a slightly different belief system.

But the overall belief system of humanity, our collective neurosis, is based on fear. It manifests in our reality as violence, crime and war. The overall resulting reality is more like a nightmare. In a society ruled by fear we see human suffering, addictions, hunger and injustice. A life or a world ruled by fear is hell.

We are not going to hell, we are already there. We are living it. It is time for a new set of beliefs, a quantum paradigm shift.

In the beginning:

There was no beginning. There is no end – the very definition of infinity. There was no god who created the Universe if there is no beginning. It would be impossible, just as it is impossible for there to be an end. The space/time continuum is an infinite loop.

Now that we have disposed of the concept of a god, where does that leave us? Well, to answer that we would have to take a different approach.

Our brains work much like a computer and like a computer, they are programmable. We are born fresh, unprogrammed and innocent. We have not yet partaken of the fruit of knowledge. As soon as we arrive, our programming begins. Those who have arrived before us and have, to varying degrees, already been programmed, indoctrinate us into our respective cultures. It is like installing an operating system on a computer.

This is what has been termed the Matrix. The program predisposes our views of reality. This program has taken many thousands of years to develop. Everything is neatly packaged, categorized, named and divided into compartments. There are those among us who, for whatever reason or fluke of events, have escaped the Matrix or at least have seen it for what it is.

Many ancient cultures also had those who saw through the veil. Today we call them shaman but that is an inaccurate term. In some cultures there were those who studied and meditated all their lives to disengage from the Matrix. They were called adepts. No matter the methods, the end result was disengagement from the cultural programming.

What one finds there is a whole complex reality, far and beyond the Matrix with infinite possibilities and potential.

You life only really begins when you unplug from the Matrix. Up until that point, you are only acting within the confines of your social conditioning.

You came to this planet, at this time, to break free of the mind numbing restrictions place on you by society and to help others do the same. Did you think you were born for anything less?

You chose to be born at this time because humanity is facing the most dangerous time in history: the total annihilation of our sovereignty and freedom. You chose to come here to prevent this from happening. It is up to you to warrior up and take a stand.

Spirit

We talk about the spirit realms as if they are some wondrous and mysterious place beyond our grasp and elevate those who visit there to a higher status. But what is spirit?

Spirit is conscious energy that can take on any form to suit the situation or the beliefs of the observer. That is why some people see angels, some aliens, some animals or plants, etc.

At our core we are conscious energy that has taken the form of a human for the purpose of educating ourselves in the truth of our reality, to realize who and what we truly are. And we are so far beyond what we can imagine with the restricted view we have been programmed to accept.

When you look at a solar system, you see almost the same configuration as an atom. If you look at a galaxy you see almost the same configuration as a molecule. One is the microcosm and the other the macrocosm. As above so below, like a Fractal that replicates itself as you zoom in all the way to infinity.

https://www.youtube.com/watch?v=L0roeFpanSl

In the quantum world, there is no matter without an observer. You are that observer and everything you witness and experience is dependent on you to exist. Your existence is also dependent on an observer and it is dependent on an observer, ad infinitum. We are one spot on the fractal and yet, we are the fractal.

The point is, who or what creates this Universe is just as much a part of you as you are of it. If you want to call this conscious energy gestalt "god" then it stands to reason that you are also "god". I prefer to view it as The All That Is. There truly is no separation. We are all One, just that most of us have not realized it yet because most are still stuck in the Matrix. Their ignorance of who they really are is causing a disturbance in the Force (Source) and manifests as destruction.

Words, Thoughts and Prayers

Words have power. Thoughts have power. And prayers have power. Where does that power come from? From god? If so, from whose god? From the Great Spirit? What is the Great Spirit?

What the heck am I babbling about now?

There is no god separate from Creation. Creation is god, therefore, the Universe, which encompasses all of Creation, is god, the Great Spirit. The Universe is infinite. Do we really believe we can conceive of the infinite with our finite minds? I don't think so. Does the Universe have a gender? No. Is the Universe our father? How can we even contemplate that question? The Universe may have created itself and we may be part of it, but let's put that thought into perspective.

Our solar system is similar to the structure of an atom. Our planet is but a proton circling a nucleus (the sun) and we are but quarks moving about on that proton. If the Universe is the body of the Great Spirit, then, do you really think that the Universe is even aware of this planet or the quarks that move about on it? Are you aware of the quarks that move about on a proton in your body? Not very likely. So how do you expect that your words, thoughts and prayers aimed at the Great Spirit will ever get to Its attention?

Some native traditions refer to the Great Spirit as the Great Mystery. Why do you think that is? In the old traditions, the Great Mystery was never referred to as Father. It was understood that it was an unknowable entity too vast for our finite minds to comprehend. And that is my point. The power of our words, thoughts and prayers are wasted when aimed at the unknown, the unknowable.

So, to whom do we send our prayers if we want them to be truly effective?

Science has proven that the Earth, our Mother, is a living organism. We do not live on our Mother, but are but a single cell of that organism. The Mother is the source of all life on this planet. The energy field we call the biosphere is the source of life. The energy that is generated by our words, thoughts and prayers owe their power to that energy field. Thus it only makes sense that we direct our prayers to Mother Earth.

Actually, I doubt the Earth has a gender either, but since It is the Source of life here, we refer to It as Mother. It is a tangible part of our lives and all our experiences are part of the life experiences of the planet. It just doesn't make sense to me to pray to the intangible, the unknowable, when the tangible is right here. The power behind our words, the Source of our thoughts, is the biosphere.

We are all One

I was thinking about how that works. I came up with an analogy of the compound eye; we are each a unique perspective of the Human Race like one of the compound eyes. Because the eye is convex, each individual eye has a slightly different perspective and the combined images produce an overall image that the brain perceives as a single image. I think the process is similar. We grow and evolve as a species with the combined data of all the different perspectives (individual data accumulated) as each individual evolves. This combined evolution is how the Human Race learns through our individual learning.

The Human race, as a whole, is just one of many perspectives of all the different life forms on Earth, with each species having its own compound eye made up of all the different perspectives of each individual of that species. And each compound eye of each species would be one eye of the compound eye of the biosphere.

Each planet would be one eye of each galaxy, and again, each galaxy in the Universe would have a similar makeup of compound eyes, that makeup the ultimate compound eye of the Universe.

It seems overwhelming to think of the complexity of the trillions and trillions of individual eyes that make up the Universe, but I think it is sufficient to say that we probably should only concern our selves with our responsibility to the Human Race as a whole and our role in the overall health of the biosphere that we are an intrinsic part of. The rest of the system can take care of itself.

Where do we go from here?

The Rebirth of Mother Earth is happening before our very eyes. I wasn't sure that I would see this day (well, at least in this physical body). The quantum leap into the void has begun; the new paradigm is being birthed as we speak. This is a momentous occasion that requires great and jubilant celebration... but we are distracted by the vested interests who would prefer that we did not wake up; that we stayed asleep in the Matrix.

We are distracted by the fear mongering, the psycho-drama being played out in the media, the circus of the absurd being played out in the US elections, wars, starvation, violence in the streets. The media focuses on this stuff to keep us distracted and in fear. But it is that focus that perpetuates all the negative stuff going on. On a cosmic level we create it by our focus.

If we want peace, focus on Peace not war. If you want to end violence, focus on Love. If you want Unity, then remove your attention from that which tries to separates us. If you are feeling isolated, then it is time to look for kindred spirits and gather, share food, wisdom, connect with heart wide open.

It is time to gather. Any reason or pretense will do. Organize a dance, a sew in, a potluck meal, a sweat, a drumming session; pick each other's cooties like a bunch of chimps. It doesn't matter what you do, it matters only that you connect.

We have been hiding (to some degree) in our shells, living in an induced state of fear artificially created by a concerted effort by the ruling elite to keep us from realizing our divinity, our collective fate and destiny.

"We are stardust. We are golden. And we got to get ourselves back to the garden". That garden is in your heart and your soul. Together we can move this vehicle forward and avoid the train wreck that the ruling elite would have us believe is inevitable. It isn't. We are on the cusp of that giant leap forward we have all been waiting for. Let us hold hands and make that leap together.

All my relations.

The Reality of Our Situation

I've always said that to the ruling elite and the government, we are just chattel. Canada and the US are corporations, bought and sold on the stock market and we are assets of those corporations. Stop expecting the governments to care about us. They don't. Trump has just brought that out into the open but it has always been thus. Yes, we are human capital stock. Yes get angry. Now channel that anger into constructive action. We do not live in democracies or republics. We live in corporatocracies/oligarchies. Time to change that. Time for people to wake up to reality.

The self righteous

There is a special place in hell for the self righteous. "No man is an island". That also applies to religions. Until everybody realizes that we are all in this together, that there is no separation, there will never be peace on Earth. Your god is no better than anybody else's god. Until this whole, "my god is bigger than, more mighty than your god" shit is laid to rest; there will always be war.

Pandemic?

If you are getting stir crazy, being cooped up during this self isolating stuff, just think of all the animals, plants and birds that are getting a break from all the pollution, murder and mayhem caused by humans. That is how I get through this insanity. Yup, humans will die and suffer but we are not the number one species as we have been lead to believe. We are just another animal on this planet. Humans have been wreaking havoc for too long and Karma is a bitch but the planet, Mother Earth, needs a break from us. Just keep in mind, that this is best for the overall well being of all our relations.

Consumerism

Most people are trying to be who their parents and peers want them to be and have no idea who they are or what they want out of life. They try to fill the void at the center of their being with stuff because they have no spiritual center, no grounding in reality. They are so caught up in this mindless consumer society. People who live close to the Earth do not need consumerism. Their center is full.

Control

What cracks me up is that Canada is not even a real country yet we act like we are a legitimate colonial country, dictating what the original inhabitants can and cannot do. The indigenous people have less than 1% of the original land mass and every time the corporations decide they need more land to rape and pillage, they always take from the indigenous people. Canada is a corporation, the RCMP are corporate thugs, our so called government is not really a government but a board of directors and corporations own the "government". Canada is a tax farm and its "citizens" are nothing more than chattel of the company. We are traded on the New York Stock exchange. The sooner people realize that they not really citizens and stop paying into this charade, the sooner we can take away their control.

Everybody is lying to you.

The real facts are that everybody is lying to you, everybody has an agenda. The truth is only relevant to the beholder; everything else is just an opinion. People will believe only that which fits into their preconceived ideas about reality.

My preconceived ideas about reality are that the media, the government and all established medical, financial and religious institutions are run by immature, emotionally stunted adolescents who are living in a lie of biblical proportions. In fact, we live in the Matrix: a computer generated program that we have been indoctrinated into since birth and our main objective for coming to this insane asylum is to escape the Matrix and realize that we are, first and foremost, conscious energy manifest in human form. We are divine beings trapped in the lie of material reality. As long as we buy into fear we are all royally fucked.

The Great Transition

We are in transition from the old paradigm to a new one. We have not been given a manual or a plan. We have to make it up as we go. We must envision what we want, with strong intent and integrity.

What is going on, what most people focus on, is the dying of the old paradigm. If we watch it die, we run the risk of being caught up in its death throes. Focus on where you want to go, not where we have been. Let the old paradigm die of its own accord. Build your vision with Love.

Losing Your Mind

Most people are not aware that they are not free, not aware that they are not who they pretend to be. Most people are too busy living up to everybody else's expectations that they don't have a clue who they really are. They hate their job and they hate their life but they are "responsible" people so they keep getting up in the morning and going to work, paying the bills and doing what is expected of them.

The programming we operate by runs deep. We are thoroughly indoctrinated into it, thoroughly programmed. But the operating system is full of bugs. It does not serve us in our quest for the truth. How does one de-program themselves? How do we install a new operating system?

Fortunately, we do not have to wade into unchartered waters. Many have come before us and have been able to see through the veil of lies that imprisons humanity. But first we need to know what our core beliefs are that keep us stuck in this "normal" reality of fear and suffering.

Who is to blame for this colossal screw-up? Who do we blame? There is no one to blame. Everybody has bought into the lie. Our parents, teachers and leaders are all playing with the same handicaps we are. They only did what they were taught to do, what they thought was right and so do we. The only thing that makes sense at this point is to break the pattern and change the program. We can't do it for anybody else so we have to focus on ourselves.

Because we are living is a lie, life on this planet, our own existence, is in peril. We mindlessly consume the land and its resources in a vain attempt to fill the emptiness inside. Deep down in our psyche we suspect something is amiss, but are unwilling or unable to acknowledge, that the reality we bought into is false. We have our opinions and beliefs handed to us by the media and experts. We have forgotten that life is an opportunity to grow and learn from our experiences.

We cannot know anything that we have not personally experienced. Do not take anybody else's word for anything. Everybody has an agenda. Everybody puts his or her own spin on the truth. It is only their opinion and is not necessarily the truth for you. Life can only reveal its secrets by our own hands on experience. Nobody can do that for us. Reading other people's opinions, watching TV or listening to other people talk is a complete avoidance of personal experience; a diversion.

We have been indoctrinated into the myth of entertainment as activity. Our cultural icons are entertainers, including religious and political icons. If you have any doubt about it, think about the gong show of the US elections. People were sucked into believing it had any validity but it was highly entertaining if one could stay detached from the shear grandiose spectacle of it.

The Native Americans said that your world encompasses the distance you can walk in one day. That is your realm of experience and influence. Everything beyond that is just distraction. Focusing on what is beyond your realm of influence keeps you from being here and now in the present.

Most people live their lives vicariously through others. They read magazines and watch shows that follow every movement of famous people instead of getting a life of their own. The media is a tool of the ruling class. It is designed to distract from the truth and self-awareness by entertaining us with irrelevant nonsense. Perhaps it is time to shoot your TV.

We are the aliens
we have been waiting for.

Many believe the aliens will save us from our own stupidity (environmental destruction) but I think that we have been destroying our environment because we have no connection to the Earth, at least the European descendants don't seem to. Capitalism is our invention and is the main driving engine of environmental degradation.

The way I see it, we are the aliens to this planet. At least, we don't seem to have a connection to life on the Earth or to the Earth itself. We are so disconnected, that we feel no empathy with the life forms we destroy in our quest to fill the void at the center of our being.

We keep consuming stuff we don't need. The problem may have something to do with the alien DNA that was spliced into our own DNA, a long time ago, when homo sapiens evolved suddenly from homo erectus. Why else would we feel so disconnected from the rest of life?

In a conversation I had with a friend...

She: So tell me about the matrix... are each of us just making this shit up as we go? How else could people see things so very differently?

Me: The Matrix is a program that we were all initiated in from birth. We are just repeating a pattern of thought that we inherited from similarly initiated people. To break the cycle, to escape the Matrix is our life's goal. We cannot reach our full potential as divine beings as long as we are stuck in the illusion. There have been many before us who have escaped. They tried to tell us how they did it but most are too stuck (asleep) to even know they are asleep or stuck.

She: So we're usually just jumping from program to program thinking we are making progress.

Me: Pretty much. The biggest hurdle to over come is the Ego. It will trick you into thinking you are making progress but it is only trying to protect itself from destruction. Most gurus and medicine people I have met have been successful at killing the Ego. At this point, all we can do is get it under control enough to plan our escape.

Every tribal society ever had an initiation process that killed off the childhood ego so that the person could be born into its adult duties within that culture. We have no such process in this mindless consumer society. If you know of a medicine person who can perform an initiation rite, you may be able to join them. Otherwise, at best, we can have one foot in this reality and one in the other. I figure, if I wake up in the morning and I am still here, I haven't made it yet.

She: Do you believe the earth is ascending?

Me: I hope and pray we are ascending but sometimes it is hard to hold that thought. When I see so many people who are so stuck that they believe everything the government and media tell them, I lose faith sometimes.

She: I guess if the earth is ascending it's just another realm from which we create... it's not the end game.

Me: Ascension is a never-ending process. The Universe is alive and in a constant state of growth, just like us. We are the Universe, after all or at least, a microcosmic reflection.

Moderation in all things...

I used to live in the bush. My diet consisted mostly of what I foraged from the bush. You can't get much more organic than that.

When I was in my early thirties, I was a self righteous purist. When working for the Department of Holidays (Highways) in Quesnel I would loudly proclaim my disgust at the other employees' habit of consuming vast quantities of junk food. One day, one of those employees became tired of my self righteous attitude and said, "You and your kind are a dying breed. Do you see this Mac burger I am eating? This burger is full of chemicals that are helping me to create new forms of disease that you purists will have no defenses for. We junk food junkies are going to wipe you out with these new diseases!"

Four years later we started to hear about AIDS. Two years after that my brother's doctor told him "McDonald's (the symbol for junk food) causes AIDS because the chemicals in it are breaking down our immune systems." Well, blow me away, if that co-worker wasn't a prophet after all! (By the way, I now consume some junk food just to be on the safe side.)

Then one day I was hit by a logging truck while riding on a snow mobile. I realized that no matter how healthy you are, you are going to die anyway and probably from something that is not even related to your healthy diet.

Enjoy what pleases you. You will not get any brownie points for depriving yourself of things you love. The Universe doesn't give a shit what you eat or think or believe. In the end, our bodies are just worm food. Moderation in all things folks. Fanaticism of any kind is bad for you.

The Nuclear Family

The nuclear family is unnatural and dysfunctional: 1 father, 1 mother and the kids. Humans have been tribal since we climbed down out of the trees. It takes a whole tribe to raise children.

The industrial revolution was the beginning of the end for human dignity. It creates dysfunctional people and those dysfunctional people create dysfunctional institutions (political, religious, social and economical). We have got to get back to tribal living or go the way of the Dodo.

The Earth Mother will find a way to return itself to balance and only those who align with her energy will survive.

What is Love?

Love is a verb and a noun. It is the foundation of all that is. The entire Universe and everything in it is Love. You are either connected to it or you choose not to be. You are either in a state of Love or you block it from your life.

You cannot be in love with someone. You can be In Love with Life. You can be In Love within yourself. And you can share that Love with others. If you are in a state of Love, others will want to share it with you. But you cannot be in Love with someone. Romantic Love is a myth.

If they are both in a state of Love and you share Love with each other, that is pure and sacred. But if one of you is not in a state of Love, it is too one sided and is not pure or sacred.

What most people call "falling in love" is just hormones and neediness. It really has nothing to do with Love. Sometimes, people can live together for a lifetime and never be in Love and sometimes they can grow into Love. It takes two to tango.

We each have that choice to take. We can grow to love ourselves and we can find another who is in love within them self. That is the goal: two people in a state of Love sharing their lives in a state of Love or growing in Love within themselves and then sharing with each other while in relationship.

Riding Life's Wave

Life is such an amazing ride: up, down and all around. The trick is to see beyond the surface of every event to the lessons contained in each experience, no matter how painful or out of control it may seem to be.

In the summer of '96 I had two heart attacks. OOOOW. Scary right? Wrong! Sure it was Life slamming me upside the head with a two by four but I really needed it. Sometimes I get so lost in my head I forget to pay attention to Life's little signals that something is out of balance. Sometimes a rude wake-up call is in order.

I was really feeling sorry for myself because I found myself living in a tent with no job and no money. (Remember, I ain't no spring chicken, anymore!) I was so focused on the negative aspects of my situation that it just kept on getting worse. Eventually, this preoccupation with negativity caused my body to implode. I was sitting in a chair one day and I was unable to move. My chest felt so constricted, it was like the life force was being squeezed out of me.

"Hey! Wait a Minute! I'm not ready to go yet" I shouted at whoever was squeezing me. Then I realized it was me! I had given up hope. I chose to wallow in self-pity. I was using all this negative emotion to feel trapped instead of using it as an impetus to propel me out of the situation.

When I realized I was dying I suddenly remembered there were a lot of things I came to this planet to accomplish. My life's purpose had not been fulfilled (not that I had any clear idea what that meant). There were a lot of places to go, people to meet and things to do. So I called out to the universe for help. Through a series of events I ended up in the Nelson Hospital ER being brought back from the brink by complete strangers who were treating me with more respect, concern and dignity than I had allowed myself.

The next morning I woke up in intensive care - laughing. The duty nurse came over to inquire why. I told her I had just remembered the scene in the ER. I had been completely hysterical. There was a group of staff hooking me up to machines, taking my blood pressure and sticking needles in me while I screamed in pain. One nurse was trying, without much success, to insert a shunt in my right wrist. Each time she missed I would scream at her because it would cause my heart to contract. After she finally got it in she grabbed another and approached my left arm to insert it.

I sat up right and yelled at her "No more *#@%*#@!! needles!". The doctor, a tall, slim man, stood calmly by my bed with his arms folded across his chest. He leaned forward and in a very quiet, soft voice said "Now Mister Woffenden, you are having a heart attack. You must relax and let the nurse do her job."

"No Shit!" I yelled at him. "Easy for you to say. You are not the one dying here!" I thought to myself. I continued to freak out until they knocked me out with morphine.

It seemed quite hysterical to me the next day when I thought of how calmly that doctor said I was having a heart attack; like I didn't know already! I was also laughing because it occurred to me that I must have evolved - I didn't need to get run over by a logging truck this time. (In 1980 I was struck by a logging truck while riding a snowmobile - but that is another story.)

Once I was stabilized, I was flown to Vancouver's St. Paul's Hospital for an angiogram. My test indicated a 95% blockage in three of my arterial arteries. The doctors recommended immediate bypass surgery. The thought of having my chest ripped open was more traumatic than the heart attack. I adamantly refused surgery.

At one point I had about ten doctors and nurses standing around my bed arguing and badgering me about my refusal. They told me I was going to die within a month if I didn't let them cut me open.

Don't get me wrong. I am most grateful to all those doctors. Indirectly they gave me the strength of will and the determination to walk out of that hospital and take back control over my life. There is nothing like having someone with a scalpel, trying to cut you open, to inspire you to regain your appreciation of life.

I told those doctors that I was going home to "fix" myself. They told me a person who has himself or herself as a physician, has a fool for a doctor. I told them I was aware this was their belief but it was not mine and that was all that mattered to me. I may be a fool but I am a live (and scarless) fool. I'm relatively happy, healthy and glad to still be on this planet - as weird as it may be.

So, how come I'm still here? I did a lot of research and found a diet that I figured would clear the plaque from my arteries. I took vitamin E, Lecithin, Hemp oil and rhubarb. I spent two months totally focused on my health. I walked several miles a day and took a vitamin B complex stress formula to help me stay calm.

What did I learn from all this? I created the whole drama by my focus on the negative. I had neglected my responsibility. I got out of life what I was putting into it. I was not paying attention to my thoughts and words. Consequently, I created a chaotic reality for myself. I was not a victim of any outside influences or cholesterol in my arteries. Silly, but that is the way life works.

Now I'm learning to be more aware of my mental and emotional processes. I'm more careful of what I'm putting out into the universe. I'm learning to rediscover my sense of detached amusement about this wonderful psychodrama we call life on Earth.

Here is a little story for y'all about life:

The first time I went to the Morley gathering [1978; Stoney Reserve], I spent the first three days wandering about aimlessly, not being able to make any connections or find any natives to talk to me, let alone teach me anything. I met several other 'white' people who were complaining about the same thing. One was a young man from northern Manitoba named Dave. We would bump into each other occasionally and ask each other if we had found something - a teacher or an event to participate in. We agreed to let the other know if we did.

Then, early on the fourth afternoon, I spotted some native men working on something by the edge of a clearing at the far end of the camp. I walked over and inquired if they needed any assistance and they said they would welcome it. As it turned out, they were building a sweat lodge.

This was a turning point - lesson number one.

Now that I was not looking for what I could get out of it but offering to help, things started to come to me. These men taught me how to build a sweat lodge and the fire to heat the rocks in their traditional way. When we were finished with the preparations for the ceremony, I was invited to return later to assist in it. I left feeling elated.

On my way back to camp I saw a group of people sitting in a circle at the edge of a ridge. They were listening to a woman talking. I sat down in the grass about fifty feet away, listening but pretending I wasn't. I was acting cool and disinterested but in fact I was feeling shy and unworthy of her knowledge.

What intrigued me was her ability to synthesize religious and spiritual knowledge from all over the world into one smooth flowing dialogue, as if they all came from the same source - second insight!

When she finished and everybody was leaving, she strolled right up to me, tapped me on the shoulder, looked me straight in the eye and said, "I think you are the person I was to contact today to give a message. I have no time right now but if you come to that tipi" (pointing to it) "at 7:30 tonight we will have more time to talk." Then she strolled away, leaving me sitting there feeling like a complete dork.

Eventually I bumped into Dave and told him what had happened, about being invited to a sweat lodge and the meeting with this woman afterwards. He had a completely uneventful day and was feeling quite depressed. When he heard my story he excitedly asked where the meeting was and if I minded if he tagged along. For some reason I was compelled to say that if he was meant to be there he would find a way to do so. I left him standing there looking very dejected.

Later that night, sitting in the tipi, after finding out the Red Tailed Hawk was my spirit guide, I heard the door flap open. I turned to see Dave with his eyes as big as saucers and a smile from ear to ear. He was beaming like a halogen lamp saying "I found it!" I smiled back and beckoned him to sit beside me. I asked how he found the place.

He said he was wandering about in the dark, completely lost, when he tripped on a smoke flap pole from a tipi a little ways away. He lost his balance and lunged into some guy who was having a pee out back of the tipi. They both crashed to the ground in a heap. The other man yelled "What the hell are you doing?" Dave replied "I'm looking for knowledge." The man said "For crying out loud, it's the third tipi on the left."

Life is kind of funny that way. Just when you figure you're lost in the dark, some stranger tells you where to go.

The Truth?

The truth is only relevant to the beholder. Your truth is not necessarily relevant to anybody else. It is pure ego that thinks their truth is the only truth. There are no universal truths. The moment your truth leaves your lips or your fingertips, it is only opinion to everybody else.

Each individual has different life experiences, beliefs, life circumstances, upbringings, etc. It is those things that colour your beliefs about and perceptions of reality. No two people are the same and neither are their perceptions. Belief is not truth. It is only your opinion. The Universe doesn't care what you believe, think or say. It is non-judgmental. We are just atoms in a much vaster organism of the all that is. What that is is beyond our finite minds to comprehend.

The Healing Power of a Native Gathering

This is a story of my first native gathering and the healing power that can be generated by people coming together to heal:

The magnitude of the energy that can be used when a large group of people is gathered was dramatically demonstrated to me while attending the Morley conference in 1978. On the fourth night of the gathering, of over four thousand participants, traditional drumming and dancing was scheduled. I was excited by the prospect of attending. I had never been to a Pow Wow before.

Shortly after dark, I heard the singing and drumming start some distance from my camp. As I set out in the direction of the music, I looked up at the sky. It was very clear and the stars were many and bright. However, when I was half way to the location of the Pow Wow, I noticed the air was beginning to vibrate causing the stars to blur. Soon I saw the stars were obliterated altogether. This development only fueled my curiosity.

About three quarters of the way to my destination and about a hundred yards to my left, up a small rise, was the site of the sacred fire. When I reached this point I stopped in my tracks. I was overwhelmed by a strong urge to go to the fire. I tried to ignore my inner voice. I remembered being told earlier that day, by one of the elders, to always heed this voice.

Eventually I gave in, aborting my original mission and went to the fire. Upon arrival I saw an old Navaho Medicine man, his wife and a young couple, each holding a child on their laps, sitting opposite each other around the fire. The old man was laying out his ceremonial paraphernalia in obvious preparation to perform traditional healing. I sat on the ground a short distance from them wondering why I had been summoned. The old man obviously knew what he was doing but I had no experience in such matters. I was determined to find out though.

Soon the preparations were complete. The old man turned to me and said that if I did not intend to stay until the end of his ceremony he would request I leave before he began. Leaving during the proceedings would disturb what he was trying to accomplish. I answered that I understood and would like to remain, if it was all right with him. He thanked me for staying, which puzzled me. I also had no idea why or of what use I could possibly be. I was soon to find out.

Before he began, the voice in my head said I was not to watch because this was his Medicine and not for my eyes to see. I was sitting on the ground with my legs folded in front of me, so I closed my eyes and let my head drop forward. Almost immediately I heard a thunderous roar in my ears, like a rocket taking off and a warm pressure at the centre of the top of my head. The pressure and noise increased until my head seemed to open up and a beam of light poured out. As the light grew brighter I 'saw' the beam go out across the fire and split in two. Each part of the beam then entered the foreheads of the two children.

None of this did I will to happen, nor did I understand what was happening. The experience was so intense that I was unable to think at all. This light was issuing forth from my head with such force that it was difficult to prevent myself from being pushed over backwards. It took considerable physical effort to hold myself steady so as not to deflect the beams away from their targets. I began to shake and sweat from the strain.

After what seemed like a very long time, the intensity of the light and noise began to fade, then stopped. I could no longer 'see' the children. Slowly I raised my head and opened my eyes but they fell shut instantly. I had only a brief glimpse of the scene before a new beam of light burst from my forehead accompanied by the roar of rocket engines, and went out across the fire as before.

This time I found it more difficult to resist the backward push on my body. I was becoming exhausted and my clothes were damp with perspiration. I was beginning to get chilled in the cool mountain air. The pressure and intensity would not subsist. All I could do was hold on. After what seemed like an eternity, the light and noise stopped and I collapsed on my side, panting and sweating, too exhausted to open my eyes.

I must have passed out or fallen asleep, for when I finally did open my eyes, I was alone. I had heard no one leave. I was too tired and cold to think or care about anything. I just wanted the warmth of my sleeping bag, which I had to crawl a quarter of a mile or so to get to.

The next day, I awoke with a clear understanding of the preceding evening's events. I had been used by the old shaman as a channel for the energy that had been concentrated in the area by the drummers and dancers. The energy had flowed through me in a focused form and passed on to the children. The ordeal was so hard on me because of the intensity of the energy harnessed and because I was not used to being used as a channel.

"The Kingdom of God is within"

We are far more than just our physical bodies. At the core of our being is the creative energy that permeates and connects us to all of creation. It is the force that many refer to as God. It is only through this connection to the creative force that we may learn the truth of who we are, our purpose for being here, and where we came from. All paths to the truth, point inward to our Selves. There are no answers to be found outside ourselves, only signposts pointing the way home.

Symbols.

The Truth, God or Reality, do not change with the symbols that we use to represent them. There are thousands of spiritual symbols including vocal, pictorial and written. The bible, ceremonial pipe, pentagrams, crosses, signs of the zodiac, runes and circles are just symbols to which we attach meanings. What people tend to forget is that words are also symbols.

Words are just a series of lines and curves or sounds. They are only symbols that we use to represent feelings, ideas and things. It is only when the eyes and ears perceive them and the mind interprets the message, relating it to past and present experience, that we attach meanings to them. This process is so fast the brain tricks itself into thinking the meaning is in the word, not in its own interpretation of it.

Words have a certain mystique. We place scholars on a pedestal and lavish awards on those most skilled at manipulating words while frowning on the illiterate. We are so enchanted with the power we feel in words we measure our success as a civilization on our ability to use them. We give names to everything we see and invent new ones to help us classify our experiences of life.

We cannot, however, experience a frog or a sunset by reading someone else's account of them, no matter how skillfully written.

Heart: Hart - a hollow muscular organ that by rhythmic contractions keeps the blood circulating in the body.

This collection of words barely begins to convey what a heart is. Only after experiencing a heart with all five physical senses do we begin to have any understanding of what a heart is. All the words ever written about heart can only give us symbols but we will never know heart because only heart can know what it is.

Words are only symbols that we use to communicate. They are not nor do they contain any reality. They, therefore, cannot be sacred (or profane for that matter). Anyone who has experienced the beauty and wonder of nature or had a spiritually enlightening experience knows how inadequate words can be.

Words used in different contexts, historical time frames, and geographical locations often will have very different meanings. Like all symbols, words are not eternal but are flexible; bending and twisting to suit the fancy of the user and open to interpretation by the receiver. Poetry is a classic example of how words can be used to create new imagery.

People tend to embrace a particular religion or philosophy based on which set of symbols are easiest to assimilate into their particular system of understanding. What most fail to see is that no matter which set of symbols they choose they are only fingers pointing the way home. The problem does not lie with the various teachings but that we mistake the symbols for the truth. Instead of going in the direction that the finger is pointing we stop and suck on the finger thinking it will somehow nourish us.

We get caught up in the dogma surrounding the words. We expect the symbols to provide all the answers so we surround ourselves with others who have chosen the same symbols to justify our choice of beliefs.

All the great masters who gained wisdom and understanding did so by paying the price of following the direction to which the finger was pointing. Jesus is said to have spent forty days and nights in the desert, fasting, meditating, praying and fighting his personal demons to gain his wisdom and vision. You cannot reach wisdom by reading books or listening to sages. You have to go beyond symbols.

The Great Deception

Christianity is based on a false dialogue. The Christ entity is a fabrication based on many other enlightened beings that preceded Jesus of Nazareth. Christ is a title given to enlightened beings. Jesus may or may not have actually existed. If he did, he was just an ordinary man like everyone else. But Constantine and Paul may have fabricated his enlightened state.

In the original Aramaic texts Jesus is reported to have said, "I am a child (son) of god" and "We are all children of god." This was changed later, during the deification process manufactured by Constantine. The whole "Christ" character was taken, piecemeal, from many previous enlightened beings, also revered as gods, like Mithras, Horus, Buddha and Krishna. (Note the similarity between the sound of Christ and Krishna.)

There were 18 years of Jesus' life in the bible, from the ages of 12 and 30 when he supposedly started his missionary work that are unaccounted for. What happened during those years? According to Hindu and Buddhist accounts, he spent his time studying at the feet of the great Hindu and Buddhist masters. There are also accounts of him spending time with Zoroaster and Egyptian masters. This would account for his teachings being very similar to the teaching of Buddha and why he was given titles the same as their gods, like, Lord of Lords, Prince of Peace and many others.

Also, some of his titles were those assumed by Egyptian pharos and Constantine himself. In fact, early depictions of Jesus were actually those of Constantine.

Various religious factions of the day were fighting amongst themselves and tearing apart the Roman Empire. Constantine needed a unifying religion to save his empire and so he amalgamated many different religious ideologies into one and passed it off as the one true religion sanctioned by Rome. But the truth is, it was a fabrication with very little basis in fact.

What is God?

What does the word god represent? For most people in our society it represents an omnipotent entity that created the Universe but is not part of it, which takes a personal interest in our lives. That entity does not exist. It is pure fabrication by an immature, insecure ego that cannot come to terms with its own insignificance.

Let me put it this way:

Horton Hears a Who – Horton was an elephant who hears a voice coming from a spec of dust floating by. He eventually discovers that an entire civilization occupies this tiny spec of dust and they are calling out to see if there are any other life forms out there occupying other specs of dust. Does this sound familiar? We are the Who that Horton hears.

Horton is an entity that occupies a world with a multitude of other entities and this world is just a spec of dust to other worlds occupied by a multitude of other entities... ad infinitum. There is no way that the Who can comprehend who or what Horton is. In our terms, the Who would think that Horton was a god. But Horton did not create anything nor is he any more divine than any other entity.

Taken as a metaphor, the story parallels our own. We occupy a tiny spec of dust in an infinite sea of dust and we are trying to contact other specs of dust (SETI) to see if we are alone or not in the infinite sea. Our comprehension of the vastness of this sea is infinitesimal. We call this sea the Universe and nothing in it exists outside or separate from it. There may be other universes out there but that is a speculation that is of no use to this discussion.

Busting Down The Walls

One day a friend and I were walking down University Street In Montreal. As we approached Sherbrooke Avenue we saw about forty people standing on the corner waiting for the light to change. A commotion at the centre of the crowd caught our attention. As we drew closer we recognized a familiar bag lady pushing her face into the face of each individual around her, one at a time, and sticking out her dentures and clacking them at these harried commuters.

These people were moving back away from her, recoiling in horror. The rush hour traffic on the street was very heavy and it was difficult to move away without stepping into an oncoming vehicle or forcing someone else to do so. Soon she found herself standing in a small clearing at the centre of the crowd.

Now I must explain her attire. She had on one of those Swiss alpine-type green felt hats with feathers in them, pixie boots with curled-up pointy toes and about forty ragged skirts that were so holey and shredded it took that many to give adequate coverage. At this point she gathered up the hems of her skirts, hoisted them up over her head and began performing some suggestive pelvic tilts while shouting, "Doesn't anybody Fuck anymore?!!"

Several people risked being killed by jumping off the sidewalk into the traffic while a few others were almost trampled into the pavement as others tried to run away down the sidewalk. My friend and I fell into the grass in a fit of hysterical laughter.

Now, here was a lonely old lady, rejected by society and vice versa, who, in an act of desperation and a lapse of social grace, was reaching out to a crowd of strangers for a little attention and social comment. There was a group of commuters, who were probably strangers to themselves, being shocked back to reality in a most bizarre manner. They recoiled from a person who was being completely honest and real. Sure her methods were a little unorthodox, to say the least, but she was reaching out to touch someone and without a telephone at that, only to find there was nobody home.

I often think of this lady and my friend, who is no longer among us, and the implications of that day. Too often we are stuck inside our shells, silently screaming in our isolation. Locked behind our protective walls, we wonder why our spouses ask "Why don't you talk to me anymore?" Most of us walk a fine line between being lost and alone behind those walls, and trying to be 'normal'. Somehow we managing to (at least to some degree) function in the outside world. How did we get here, and where do we go from here?

The last time I saw that woman was in a laundromat washing two garbage bags full of used Kleenex. She approached my friend and asked him to walk her home because the children on her street taunted her and threw stuff at her. He walked her home.

I shied away. I couldn't face that part of myself. I have regretted it ever since.

My friend used to say, "Life is but a joke and the joke is on you". One day he couldn't take the joke anymore and died of an overdose of cocaine. After That, I ran away from the city and hid in the woods, near Quesnel, for ten years.

It seems most of us live lives of quiet desperation. We become comfortably numb to the suffering around us. The space that separates us from the next person is filled with walls of our own making. There is a very fine line separating us from that bag lady. "But for the grace of God go all of us."

As a society, we have become soulless, spiritless automatons consuming mountains of material goods trying to fill the emptiness we feel inside. I doubt this was the intention of the Creative Force that caused all this magnificent beauty we call Earth to come into being. Our civilization is destroying that beauty and replacing it with plastic, concrete and pavement. It may be time to become uncivilized. Free the Earth of our civilized bonds. Free ourselves of

82

those same bonds because those ribbons of highway, those fences and walls by which we have bound the Earth are the same as the chains with which we have imprisoned our souls.

Awaken to your freedom and TEAR DOWN THE WALLS.

Bandwagons

Ya know, back in the 60s I jumped on all the bandwagons: female equality, black equality, natives, Jewish, etc. Then one day I woke up and thought "who are these people wanting to be equal to? White males? We are some of the most fucked up assholes on the planet!" Ya so, we are, but why is that?

Because we have to live up to unrealistic social expectations just like women. We all have the same emotional problems because of our social conditioning. Is it a male or female problem. No. It is a social problem. Are drug addicts bad people? Is anybody a bad person because they can't handle the stress of trying to be someone they are not? These problems are much deeper than what is seen on the surface.

We are tribal beings. We have been for a million years or so. Most tribes existed as extended families. Then, 200 years ago, along came the industrial revolution and we were forced into nuclear family situations, which are unnatural to us. The result is that everybody is emotionally stunted. Not only that but all our institutions are designed and operated by emotionally stunted people. So how do we expect there to not be severe problems with all we meet?

Solutions, however, are not so easily come by. As long as we are stuck in this society with our nuclear families, it is going to be nearly impossible to fix. We need to rearrange our social structures to reflect our tribal instincts and I think many of the younger generation are working on that. Burning Man, Shambhala and other gatherings are very tribal in nature. I know it is a slow process, but eventually, humans will evolve back to their roots.

A Lightness of Spirit

We decided to be born at this time because we are old souls who chose to help bring in the new age of enlightenment. It's a hell of a responsibility or a terrible curse, depending how you look at it. It is time to break old patterns and cultivate new ones in line with the evolution of humanity from the fear based past to the one of love and compassion. Just remember to have fun doing it. No one can reach enlightenment by being serious. Enlightenment means "a lightness of spirit". You literally need to laugh your way there.

Over Population

It is not so much that there are too many people; it is that there is too much demand on resources. Everybody wants more than they need. Hunter gatherers never taxed the system by over hunting or gathering, now everybody wants two car in every garage, a house in the suburbs, 2.4 kids, 84"smart TVs, etc. If we could learn to be happy with what we have and stop demanding more than the Earth can provide, we might handle a few more millions of people, but as it is, we can't. The Earth just can't handle it any more. Climate change is a wake up call to moderation but not enough people are willing to do what is necessary to curtail their mindless consumerism.

Love

Love is the creative energy that holds the Universe together. What most people call falling in love is more a neediness or hormonal overload. Everything is love. It is a state of being. You are either in a state of love or you are not. It really is impossible to love someone. It is impossible to know someone, let alone love them.

We fall in love with an idea of who we think they are. Love is something we cultivate in relationship. If they are in the state of love, and you are, then you can share that state. We can only see others through the filters of our own self image. The only person that we can truly love is ourselves and when we do, someone who loves themselves will be drawn to our light.

Death

"In the end times, all will be revealed." This does not mean the end of the world. It means time will become irrelevant. Time is a human 3rd dimension construct and as we transition into the 5th dimension, there is no need for time. That is the apocalypse. I have had several near death experiences and I am here to tell you that there is nothing to fear in death. Our bodies are vehicles that our souls/spirit inhabits until it ceases to be of use. Then we discard it and move onto another one.

Truth and Farts

What are the similarities between your Truth and a Fart?
- not everybody wants to hear them
- not everybody wants to acknowledge them
- not everybody appreciates them
- not everybody feels comfortable sharing them
- not everybody wants you to share them

Woke to What?

OK, what do we do for people who do wake up? What is it that we want them to wake up from and to what? Where to begin: wake up from The Lie - that we are insignificant beings separate from each other and the Creative Source, that we are mere sacks of skin and bone fumbling through life. We are much more than that. We are pure conscious energy, the same energy that created the Universe. We are divine creatures responsible for creating our own reality within the reality of the collective consciousness of all sentient beings cohabiting with us on spaceship Earth.

We have the power to change the world we live in. There is nothing to fear but fear itself, but it is all around us perpetrated by those who would keep us in the dark so they can control us. Life really begins when you stop being afraid and accept your divine birthright. You were meant to lead the way (just like we all are). The world is waiting; all living things are waiting for us to wake up before we destroy our life support system - the biosphere, our Mother Earth. So spread the word.

Karma

A medicine man once told me, "You see all those drunken Indians out there? They are reincarnated white men who tried to wipe us out and their only way to salvation is to learn the ways of our ancestors so they can learn what it was they tried to destroy." Karma has away of balancing things out, if not in this lifetime, then in the next.

Old School

I used to hang with an old school medicine man. I noticed that he would alter his ceremonies and rituals to suite who was ever in attendance. He would make off coloured jokes during the ceremonies that would upset my catholic upbringing until I realized that even the creator of all this must have had a sense of humour - after all, it created humans didn't it?

He taught me the great cosmic joke, and it is us. Since then, I discovered quantum physics, which I realized was a bridge between science and the wisdom of the ancient mystics. Besides, spirituality is a very personal experience. It has nothing to do with religion, deities or dogma.

Comfort Zone

Your comfort zone is a slow death by attrition. There is little to no growth or movement. Keep challenging yourself to change places, occupations, friends, etc. The purpose of being here in this insane asylum is to break out of it to find the real you. The craziest people you will ever meet are the ones who think they are normal, sane and smart. That is their comfort zone and they will rot in there.

Be Here Now

Be in the moment. The future is not guaranteed. Yesterday is a bad memory. Don't worry about tomorrow. Time is a human construct and is meaningless if you live in the moment. Children are on their own timeline. They will become who they were meant to be in their time. Our society puts too much pressure to become what they want. Fuck 'em. They don't know shit. Be who you were meant to be, and leave the rest to the Great Mystery.

Two Different Worlds

There are two different worlds, that we living in, on this planet. One is mechanical and the other is natural. One is the Matrix and the other is real. We live in a lie of biblical proportions.

From the day we are born we are indoctrinated into the lie. We are programmed into thinking the mechanical world (the Matrix) is reality. We are indoctrinated by our parents, priests, teachers and peers all of whom were indoctrinated by the indoctrinated. They did not think they were doing wrong because they were indoctrinated into the Matrix an they thought that was real. It isn't.

If you go out into the jungle, you will find people living in nature. They are happy, run around almost naked, are healthy and at peace with their world. They follow natural rhythms, eat mostly organic and are in sync with their environment, even though their environment is full of deadly plants, insects and animals.

In contrast, as a friend who went to the jungle testified, the invading peoples were dressed to the nines with heavy clothing and carried weapons because they were terrified of all the things that could harm them. These are the two worlds I speak of; very different, very opposite. If we look around at what is going on in the natural world, mass extinctions, depleted wild stocks, depleted soils, poisoned earth, pandemics and dwindling natural spaces.

The Big Lie is that we are separate from Nature. We are part of it, and if we do not get back to being part of it, we will perish with all the other life forms that we are killing off in our mad rush to extinction.

Many people have awakened to our peril. They are demanding that we stop our mad rush to destroy all that is natural. There is almost no old growth forests left, wild salmon are almost extinct, woodland caribou and so many other species are on the brink. We have to put a stop to our demand on natural resources or we will be forced to go back to being hunters and gatherers.

Most people will not survive that because they have no skills to go back. People who still live in harmony with Mother Earth will have the best chance to survive. In fact, evidence is mounting that we have been here before. We have had to abandon our technology either because of mass extinction events or our civilizations self destructed. We are at a turning point. We have a choice to make: continue down this path of self destruction or we go back to living in harmony with Mother Earth.

What will you decide?

We are love/god/creator

A baby is conceived and born in Love, filled with and is Love personified. The entire Universe and everything in it has Love as the basic building block. It is impossible to not be in Love. We are made of it. What some people call being out of Love/being without god, is simply the stupid notion that we are somehow separate from god/Universe/Creator. There is no separation. We are all One with creation: god is in us and we within it, we are god and god is us.

To put it another way is that the basic building block of all of creation is energy (see quantum physics). That energy is the same whether it is in an atom or a galaxy. There is only one energy that permeates all there is, and that energy is Love/god/Creator. We are conscious energy (spirit), a microcosm of the whole, that chooses to take on physical form in order to learn, or more correctly, unlearn the lie that separates us from our own divinity, our Unity/Oneness with the All That Is. It's what some people choose to refer to as god.

We are One

To me, the Universe is the creative force. It created itself. Why the Universe would care about the human drama queens is beyond me. But everything in the Universe is connected and we do have the capacity to tap into that frequency, to understand (as far as we are capable) our connection to all. I find the separation dichotomy of modern religion, to be the source of all our angst and violence.

We are not separate, god is not separate, everything is interconnected, interdependent. What harm we do to others (human, animal, plant), we do to ourselves. We have turned paradise into a garbage heap trying to fill the emptiness inside caused by this separation schism. We are all One.

Going Home

We are first and foremost spiritual beings (etheric) inhabiting a 3 dimensional body of our own making. It is the product of many millennium of focus on our 5 physical senses while ignoring our intuition. This has been intentionally exploited by a certain group of self-appointed "rulers" to entrap humans into doing their bidding. The ascension process is not one so much about evolving into something greater as it is a process of remembering that we ARE that something greater.

We do not have souls, souls have people. The "going home" is waking from this Matrix (an actual mind program) that we have been stuck in for a million years or so. The shaman understands the Matrix and can manipulate outside of its restrictive bounds. We are headed back to that place where we are all shaman, interconnected and inter-dependent. We truly are one. At this point it is not much more than a concept that we may intellectually understand but have not yet integrated it into our heart's knowing. This is about what I refer to as "going home".

Creation vs Evolution

Why does it have to be either/or. The Universe is made up of certain stuff and everything in It, including humans, is made of the same stuff. The Universe is so absolutely huge, infinite actually, and there is no possibility for our finite minds to imagine or comprehend it. It just is. Is it intelligent? Does it have consciousness? Well, to find that answer we just have to look at what we do know, what we can see and understand. There is a certain symmetry and grace to all the chaos we witness.

If we are made of star dust and everything on this planet is made of star dust (the same stuff the Universe is made of) and that the Universe itself is consciously aware of itself, and we are consciously aware of ourselves, then, to me, it stands to reason that somehow there is a connection between all parts of the whole. In other words, we are connected to all-there-is. The Universe is made of creative energy and is creative consciousness. We are also creative consciousness.

As someone put it, "We are the Universe having a human experience." It is my contention that this creative consciousness, our finite awareness is just a microcosm of the infinite awareness of the Universe. That seems to be where some get confused: the Universe is what some refer to as god. It does not judge or punish us for thinking or believing anything. It is indifferent. To think that it cares about you, personally, would be a colossal act of narcissistic egotism.

Something as infinite as the Universe and its infinite consciousness would be impossible to comprehend. Things would evolve within the parameters of infinity. Creation/evolution are beyond our capacity to even imagine, so it is a futile debate that distracts us from what is really important - living life to its fullest, being present and not distracted by trivialities.

Aboriginal peoples worldwide sensed a connection to all life on this planet because they were not distracted by philosophical mental masturbation over topics that were beyond comprehension. They had time to develop a working relationship with life around them, to be part of the cycles of life. We, in the west, have separated ourselves from those cycles, put ourselves above them with our egotistical self-importance, thinking that the Earth was created for our needs and exploitation.

It wasn't. We are part of one gigantic living being we call the biosphere. We are not separate but just a strand in the web of life. Until we understand that whatever happens to the web happens to us, we will keep going down this road, arguing over nonsense while our life support system degenerates around us.

Does it really matter if life was created or evolved? Not really. Are we putting all life on this planet in jeopardy with our egotistical beliefs in our own self-importance? Yes. Time to awaken from the drunken stupor of religion and smell the roses before we all go the route of the Dodo.

Conflict and War

There can never be Peace or Love in the world when there are still those who insist that their way is the only way. Love is accepting everyone as your brother and sister no matter what they believe; which god, whatever sexual orientation, gender association, whether they are for or against those things you hold dear. If there is only one god that created everything, then we are all children of that god. It doesn't matter if a person calls that god Allah, Thor, Krishna, the Great Mystery. Until we can accept all beliefs come from the same Source and only vary because of time, location and culture, then we will always have conflict and war.

People don't have souls, Souls have people

Souls have people. "We are spiritual beings having a human experience."

From my experience, the soul doesn't die or dissipate, it exists as part of the whole, the Universe/god/the all that is. It takes on various forms in 3D to experience life and those are to acquire greater understanding of our own divinity, our oneness with everything. Earth is a school, of sorts, and we come here for a purpose. What that is, is for each of us to figure out. Nobody and no book can tell you what that is. It is your journey and nobody else can know why you chose to come here.

Politics

Politics is a game they play to entertain the masses while the ruling elite go about doing whatever the hell they please. The ruling elite have brainwashed us into believing us that we are free, that we live in a Democracy, that money is real. Of course, none of that is true. Democracy doesn't exist. We are living in a plutocracy/corporatocracy/Oligarchy. They give us the illusion of choice but in reality, there is no choice. They own all the parties and politicians are mere paid actors playing the roll of your representative.

- Money is printed out of thin air to enslave everybody with debt. The banks and corporations own everything.

- The police are there to protect the rich from the poor. They do some token duties to give you the illusion they care about you.

- The laws were written to keep the poor from killing the rich, nothing more.

- There are no countries that you are a citizen of. You are chattel to your corporate owners.

- Countries are corporations bought and sold on the stock market.

They are using the Covid plandemic to install their New World Order, their one world digital currency, reduce the human population by 90% and enslave the rest of the survivors... unless we can stop them now. It is up to every everyone who can still breath and think for themselves to do all we can to prevent this from happening. They are few and we are many. We are the 98% who have everything to lose if we don't act now.

Fringe Dwellers

Throughout history, there have lived those, who, for one reason or another, did not fit into the niche prepared for them in society by their parents and teachers. There are those who would have you believe the family is the cornerstone of our society. They are, of course, completely off base. Conformity is the cornerstone of our society and those who cannot, or will not, conform are pushed, or graduate to, the fringes of society.

There are junkies, misfits and malcontents on the fringes but also a small percentage who are just too conscious to accept their role as sheep. They see the glitz, the glamour and the bright lights as someone else's bad dream, leading to humanity's demise.

It has always been that way but not with the overwhelming numbers and power to destroy we have today. We have never, at least not in our known history, had the technology to, not only to lay waste the planet and everything on it but to become such slovenly, unhealthy and diseased creatures. We are on the verge of extinction, not so much from the destructive forces at our disposal but more from our complete indifference to life in general. Life has been replaced by "things" to be consumed. As one musician (Roger Waters) put it, this species has amused itself to death.

Most fringe dwellers, I have met, feel the same way about society. They just express their mistrust in different ways. Most end up conforming to the status quo of their chosen fringe group. The radical fringe dwellers, however, are few and far between because they walk alone on their chosen path. It's a lonely life, but it is the only one that makes sense to them. To stand in the center of one's own truth requires a tremendous commitment, no group affiliations (except to the species one finds oneself born into) and an absolute faith in life and the Universe. A tough job but some of them feel compelled to do it.

What motivates people is confusing to them. Early on in life, they came to feel like an alien on their home planet. Even in their adult life, there are days when they have felt they had awakened in a completely unknown environment with no points of reference. Even the scenery is unfamiliar and the creatures they are supposed to communicate with are incomprehensible. Their dialogue sounds like gibberish to them.

Since as early as they can remember, they have had difficulty relating to others. Nothing here seems to make sense. People seem to be oblivious to their nihilistic patterns of behavior. Society seems to have lost sight of the concept of a racial continuum. I call it the Hoover syndrome – you know - they suck up all the junk they can get their hands on and the one who dies with the most junk wins! There's no thought for the well being of future generations. They don't seem to realize there is only a finite amount of resources to create all this junk from. Mindless consumerism is just beyond their comprehension.

When I looked at this phenomenon, I realized that I have been a fringe dweller almost all my life. I've always felt like an alien and what most of society hold dear and true does not make any sense to me. I have lived on the fringes since the mid '60s. Society has become a flock of sheep being led to the slaughter by an elite group of sheep.

Most people are oblivious to the fact that they are brainwashed by the media, schools, religion and governments. But I see that is changing and fast. There is hope on the horizon and we are approaching that horizon at warp speed. The Great Awakening has begun and I pray that it will be soon, and that humans will finally live in peace and compassion.

Life is like a movie

Life is like a movie - your movie. You spend your time trying to find the clues that will help you find your way home. You create it by projecting your thoughts and beliefs out into a seemingly solid world. But it is not solid and it does not have a mind of its own. Life reflects back at us our thoughts and beliefs about ourselves and reality. If we change our thoughts and beliefs, we can change our reality.

It is different for those who blindly accept the status quo, however. They do not take responsibility for creating their reality, so they live in someone else's. Hell, most of them don't even know they are creating their lives by default by following someone else's script. They will complain and whine a lot but they will do very little to change it.

Most have access to this information. Many are aware of it but still they refuse to take that leap of faith because they desperately cling in fear to their old beliefs. It may be uncomfortable, even painful, but it is familiar and therefore safe.

Safety and security are the primary causes of the death of spirit. They sap the life force out of life. Without a sense of wonder and adventure about life, death of the body is inevitable.

We have been exposed to a set of beliefs about the nature of reality that says life is a certain way. There is far more to life than that particular view. There are a whole bunch of other rules by which to play this game. By learning these rules, you can take control of your life and create a life that is more suited to your liking.

Most people are stuck in the dream. Those that are aware of these rules, they think they can use them to improve their lot in the dream. They don't realize that the dream is a lie.

Nothing has been hidden. All this knowledge is available to everyone. Most don't know what to do with it. Most people think that the point of the game is to accumulate knowledge by piling up new information on top of every other bit of information they have gathered over their lifetime.

In order to live outside the lie you have to first unload all the old garbage and start with a clean slate. Applying the new rules to the old game is a complete waste of time because you are still stuck in the lie, only now you have to operate with a bunch of conflicting beliefs. It becomes so confusing that many end up turning to religion. That way they don't have to think about any of it anymore, because someone else tells them what to believe. It may simplify their lives somewhat but it certainly stunts their spiritual growth.

A terrible tribulation is coming. I have seen the control freaks gathering their forces. All those who remain stuck in the lie, will suffer;

some more than others. The control freaks, whether they are religious, political, industrial or military leaders, live in and derive their power from the lie. They have no hold outside the lie, which is where we should live, more and more each day.

Shamanism & Integrity

Shamanism is practiced by every tribal society on Earth. When I finally met a genuine shaman back in the late 70s, he did not use any kind of hallucinogen. He took me places hallucinogens couldn't. At the same time, I realized that I had been practicing shamanism on my own for a decade without knowing what I was doing. I was just following my intuition. I spent 10 years living in the forest (squatting in an abandoned cabin).

I studied psychology, metaphysics, quantum physics, parapsychology, history and native spirituality before I met my teacher. He was the closest person to embody Christ Consciousness I have ever met. What I gleaned from all that is that it doesn't matter what you believe, what language you speak, what the ceremony looks like or who you pray to. It is all about energy, intention and integrity.

Love is...

Love is not an act. It is not something that we do. What people call love is nothing of the sort. It is a longing, a need, a feeling of emptiness that wants to be filled by another. That is not love. Love is what the Universe is made of. It is an energy that permeates and binds everything. We are either in (a state of) Love or we are not. Desire comes from the false dichotomy of separation. The purpose of the spiritual quest is to recognize that we are Love eternal. The rest is just entertainment; chop wood, carry water.

The World is Waiting...

OK, what do we do for people who do wake up? What is it that we want them to wake up from and to what? Where to begin: wake up from The Lie - that we are insignificant beings separate from each other and the Creative Source, that we are mere sacks of skin and bone fumbling through life. We are much more than that. We are pure conscious energy, the same energy that created the Universe. We are divine creatures responsible for creating our own reality within the reality of the collective consciousness of all sentient beings cohabiting with us on spaceship Earth. We have the power to change the world we live in. There is nothing to fear but fear itself, but it is all around us perpetrated by those who would keep us in the dark so they can control us. Life really begins when you stop being afraid and accept your divine birthright. You were meant to lead the way (just like we all are). The world is waiting, all living things are waiting for us to wake up before we destroy our life support system - the biosphere, our Mother Earth. So spread the word.

There is nothing civilized about civilization

Every civilization has gone through the same pattern of rise, decadence and fall. And ours is in the fall stage. How did we get here and where are we going next?

During the rise faze, there is much war, conquest and plundering of other nations. As the population grows, more land is needed to feed them. To get more land, one needs an efficient military to go and take the land from others. A military does not just include those doing the fighting, It needs a large group to feed and supply those who are doing the fighting. The larger the military, the larger the supply chain and the more land is needed to grow, feed and supply them.

In the mean time, the folks back home are living rather decadently and coming up with new ways to kill and conquer more people so they can continue to live in their decadence. They produce more babies to supply the bodies to kill and be killed. Of course, the ruling class do not want their progeny to die, so they create a lower class to supply the bodies necessary to maintain the status quo. Sooner or later, the lower classes get fed up with their lot and start to revolt. We have been in the stage for over half a century. The ruling elite hatch a plan to save their asses from the revolting lower classes.

This is where the civilization begins to collapse. At this time, they have designed the Great Reset in an effort to retain their power. Their plan is to wipe out about 5 billion useless eaters and turn the survivors into cyborg slaves. Unfortunately for them, too many people are on to their plans. That is why they had to start a war in the Ukraine to divert everybody's attention away from them.

At this point we do not know the outcome of this diversionary tactic but I believe we are going to awake from this dystopian nightmare and ascend into the 5th dimension in spite of their shenanigans. We'll see. Initially, their diversion seems to work but I think there are too many aware of what is really going on for them to succeed in their plans. Maybe not but I prefer to remain hopeful.

In Contusion...

Politics is an illusion, a scam, the reality of which is completely the opposite of which most people believe it is. Democracy isn't what we have been told it is. It is more like an oligarchy owned and run by psychopaths; rich power hungry lunatics whose sole motivation is control and the accumulation of monetary wealth. They are heartless and soulless humanoid parasites feeding off the bones of a dying world, a world that is dying from their insatiable appetite for more and more. They have no empathy or sympathy for other humans who they see as means to an end. They really don't care who or how they harm or kill in the pursuit of more stuff.

The really disheartening thing is that most westerners hold these tyrants as the epitome of human evolution. They strive to become them, to emulate their greed. They hang on their every depraved word as if that is the truth, not realizing how much these scum hate the very existence of other "lowly" humans. They view the masses as chattel, a means to an end.

Mindless consumerism is the only freedom allowed these tax paying plebes who have been brainwashed by endless propaganda into believing that they are free. They are not. They are owned and enslaved by their masters, who hold them in contempt.

Religion is another form of enslavement that has been heaped upon the masses. The idea that there will be reward in the afterlife for following the rules of the psychopaths, who see the masses as toys in their depraved sexual fantasies. The last thing that those psychos want is for people to wake up and see what an elaborate hoax they have been hoodwinked into believing. If people would realize their own divinity, the jig would be up. People don't need intermediaries to connect to the divine because they already are; an integral part of the whole.

PROLOGUE

Time To Remember

I did not write this but it is too interesting to me, as it aligns with all I have learned, so I borrowed it…

A long, long time ago in a galaxy far, far away there were all of these little light beings just hanging out enjoying life in that joyful timeless dimension.

And then one day a very large, magnificent angel came to them. He had a very serious look on his face.

He was looking for volunteers for a very important cosmic mission.

"We have this small - but very special - planet out at the edge of the Alcyon galaxy called Gaia.

It is quite unique like a beautiful garden and it is teeming with hundreds of thousands of different life forms.

It has been something of an experimental station in the galaxy and it has a most interesting humanoid life form that incorporates the very highest and lowest frequencies known in the cosmos. It is in fact the very epitome of dualism.

On the one hand it is an incredibly beautiful life form and is capable of carrying the highest frequencies of love, light and joy known throughout the whole Universe.

On the other hand it is capable of carrying the densest and darkest frequencies the cosmos has ever experienced - frequencies which the rest of creation evolved beyond eons ago.

Here is the current situation. Within the domain of time, this planet goes through periodic cosmic cycles. It is now coming to the end of two major cycles - a 2,000 year long age of Pisces and the 25,000 year long cosmic year in its journey around Alcyon, the central sun of the milky way galaxy.

With the completion of this cycle, many things are coming to an end and many things are about to begin.

But most importantly, the planet is experiencing an infusion of light that is dramatically increasing it's frequency.

As during any major time of transition, there will be a certain amount of turbulence. Some of this will be geological, for Gaia Herself is a living planet and is also evolving. But much of it also involves the hominoid species that dominates the planet.

This will not be a particularly easy time for the species - especially for those who are sleeping and those who are vibrating at the lowest frequencies.

As the frequency changes it will create insecurity that in turn will create fear.

The first era of evolution on this planet was the physical era and the key word was survival.

The second era, which is now ending, was the mental era and the key word was logic.

The third era, which is now beginning, is the era of the heart and the key word is love. This is the highest frequency.

Those who currently hold the reign of power on the planet are of the old order of the physical and mental.

To the extent that they can make a graceful transition to a heart centered and divinely guided life, it will be an easy transition. To the extent that they are unable to do this, they will experience much turmoil.

So this is the current situation of Gaia. The reason I am here is to seek volunteers who would be willing to incarnate in humanoid form on the planet at this time to help make this an easy and smooth transition. We have sent prophets and teachers in the past.

Very often they were brutally persecuted or killed. In other instances they were set up as "gods" to be worshiped and these humanoids built elaborate religions and rituals around them and used these religions to control each other. They did everything except follow the simple teachings that were offered.

So this time we are trying a different approach. No more prophets, saviors and avatars that they can use to create religions. This time we are sending in thousands - actually hundreds of thousands - of ordinary light beings with only two assignments:

Stay in your heart. Regardless of what happens, stay in your heart. Remember who you are, why you are here, and what this is all about. Now that seems easy enough, right?

Unfortunately, No!

As I have said, duality has reached its peak on this planet. This species has perfected the illusion of good and evil.

The greatest challenge you will experience is to remember Who You Really Are, Why You Are Here and What This Is Really All About. When you remember, you will be able to stay in your heart, regardless of external events.

So how will you know when you are forgetting? It is easy. Watch your judgments. The moment you notice that you are in a place of judgment you will know that you have forgotten Who You Really Are, Why You Are Here, and What This Is Really All About. That will be your signal.

Now here is the challenge. Life on this planet will require a great deal of discernment - wise evaluation of what is true, what is appropriate and what is for the highest good, both for yourself and for the planet. In many

124

ways discernment is similar to judgment. However, you will know when you are in judgment and when you have moved out of your heart, when you are in a place of blame.

We know how challenging that this planet can be. We know how very real the illusions on this planet appear to be. We understand the incredible density of this dimension and the pressure you will face.

But if you survive this mission - and it is a voluntary one - you will evolve at hyper speed.

We also should say that we know that some of you who will go to this planet as starseeds, will never germinate - never awakened to the remembrance of who you really are.

Some of you will awaken and begin to shine, only to be choked down by the opinions and prevailing thought forms around you. Others will awaken and remain awake and your light will become a source of inspiration and remembrance for many.

You will incarnate all over the planet; in every culture, every race, every country, every religion.

But you will be different. You will never quite fit in. As you awaken you will realize that your true family isn't those of your own race, culture, religion, county or even your biological family. It is your cosmic family - those who have come as you have come - on assignment to assist in ways large and small in the current transition.

True brotherhood and globalization in its highest form will come only in remembering Who You Really Are, Why You Are Here, and What This Is Really All About. It will come as you return to the true temple of Divine Presence, your heart, where this remembrance takes place and from which you are called to serve the world. So, are you ready?
Good! Oh, and by the way, there are a couple of other minor things I should mention......

Because of the density, you can't operate in that dimension without a space suit.
This is a biological suit that actually changes over time. There are many things we could tell you about this but our orientation time is short so I think you can just jump in and experience it. You should be forewarned, however.

There will be a danger that if you forget who you really are, you may think you ARE your space suit instead of the fact that it is simply your vehicle in that dimension.
Once there, you will notice that there is an infinite variety of space suits and a great deal of attention given to these. However, in spite of the infinite variety, because this is a planet of duality, they all fall into two basic categories called 'genders'. Again, we really don't have time to go into this now. But you will find your

relationship with your own space suit to be most instructive and interesting.

The other little thing is this. In order to operate in that dimension, you will also receive a microchip called a 'personality.'

This is like an identity imprint that, along with your space suit, will essentially make you different from everyone else. This will allow you to participate in the hologram there - something they call 'consensus reality'. Once again, there will be a real danger that you will become so engrossed in the holographic personality dramas that you will forget who you really are and actually think that you ARE your personality.
I know it sounds rather unbelievable right now, but once you get there.....

Again, there is so much more we could tell you by way of orientation, but we think you can learn the rest experientially 'on site." The only thing that is important is to remember Who You Really Are, Why You Are Here, and What This Is Really All About.
If you can do that, everything else will work out fine.

But take note: So few really DO remember this they stand out as 'different' and others called them 'Enlightened" or 'Awakened" and similar terms.
Strange isn't it

Well, Good Luck and Bon Voyage
See You There

I Love You
Anonymous